MW01681577

InspireScience

Science Paired Read Aloud

McGraw Hill Education

ConnectED.mcgraw-hill.com

McGraw Hill Education

STEM McGraw-Hill is committed to providing instructional materials in Science, Technology, Engineering, and Mathematics (STEM) that give all students a solid foundation, one that prepares them for college and careers in the 21st century.

Send all inquiries to:
McGraw-Hill Education
8787 Orion Place
Columbus, OH 43240

ISBN: 978-0-02-136957-7
MHID: 0-02-136957-7

Printed in the United States of America.

1 2 3 4 5 6 7 8 9 EUS 20 19 18 17 16 15

Contents

Vocabulary

environment everything that surrounds living things; the natural world

Changes Around Town

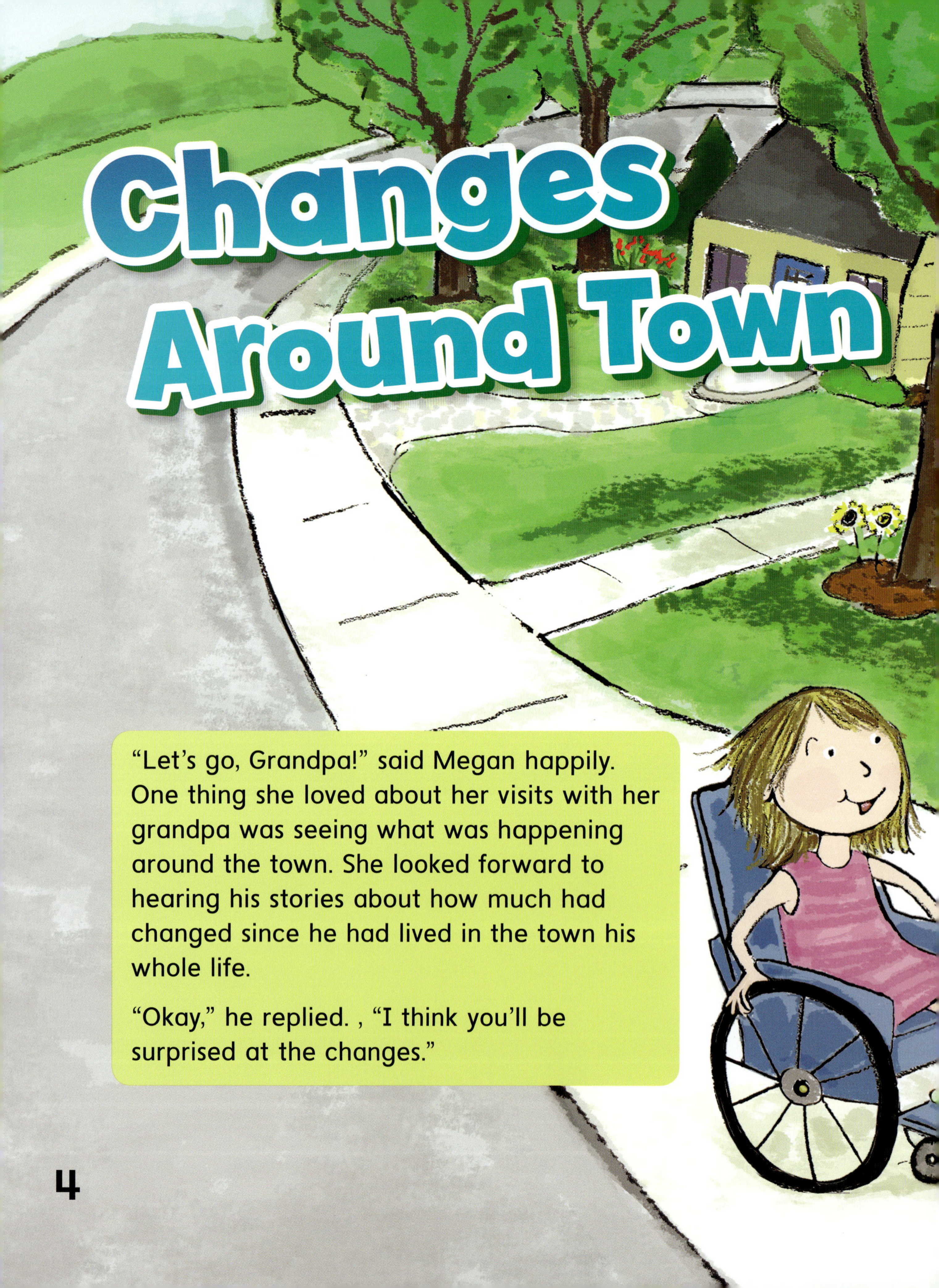

"Let's go, Grandpa!" said Megan happily. One thing she loved about her visits with her grandpa was seeing what was happening around the town. She looked forward to hearing his stories about how much had changed since he had lived in the town his whole life.

"Okay," he replied. , "I think you'll be surprised at the changes."

The first change Megan noticed was at the end of the block. She gasped. "What happened to the tree?"

The huge oak tree in Mrs. Monroe's front yard was gone.

“The roots on that tree grew down into the water pipes,” said Grandpa. “One pipe finally burst. The front yard was full of water.”

Her grandpa smiled as he continued. “But don’t worry. We all helped pay to move the tree. Now it’s in the backyard. See?”

Soon her grandpa pointed out another difference. The oldest house in town had a brand new brick chimney.

“That house has been the same for many years,” he said. “But the Cohens had to get a new chimney. Some squirrels made the old chimney their home.”

"They tried many ways to get the squirrels out, but finally they decided to just tear down the chimney."

"Where are the squirrels now?" asked Megan.

"They found a new home in a nice tree."

As they entered the town square, Megan noticed another change. Leafy vines now covered an arbor over the path. They made the spot shady and cool.

"This is a nice change," said Megan. "The shade feels good. I was getting hot!"

Her grandpa agreed. "This arbor also makes a good home for the birds." They rested for a while, and listened to the birds sing.

Megan and her grandpa arrived next at an empty lot on the edge of town. The last time Megan visited, an old gas station had been there.

“It was torn down,” her grandpa said. “I’m glad. This field is nicer than an old, empty building. The gas station had been there since I was a boy. I used to stop there for a cold drink on hot days.”

Their final stop was a favorite ice cream shop.

“This doesn’t look different,” Megan said.

Megan’s grandpa smiled. “It hasn’t always been an ice cream shop. When I was young, this was a park. The shop was built in its place. Now we can enjoy ice cream here instead!”

Plants and Animals Change Their Environments

Environments Are Important

Plants and animals need a safe home and food in order to live. Their environment is important for their health. An **environment** is everything that surrounds living things. Plants, worms, and soil are part of a bird's environment.

Birds use plants.

Chris Willig

Plants Change Their Environment

Plants can provide shade. This cools other plants, animals, and people when it is too hot.

Plants help things cool down on a hot day.

Inspirestock Inc./Alamy

Plants provide food and safe homes for animals. Where there are many plants, there are many animals.

Plants attract animals.

Plants help protect land when it rains.

Soil washes away when there are no plants.

Plants Change the Ground

Plants help protect the ground because they hold soil together. During rainstorms, soil that has plants will not wash away as fast as soil without plants.

(bkgd) Pete Ryan/National Geographic/Getty Images; (inset) Tim McCabe, USDA Natural Resources Conservation Service

Plants can break through the ground and crack sidewalks. People cut down big trees to make it easier to walk on a sidewalk. Animals that used that tree have to find a new environment.

Tree roots can break through the ground.

(l) McGraw-Hill Education; (r) woojpn/iStock/Getty Images

Animals Change Their Environment

Animals move things in their environment to make their homes. They change the land above and below the ground.

Sea turtles nest in the sand.

An armadillo builds a burrow in the side of a hill.

Armadillos make homes called burrows. A burrow is a hole in the ground. Sometimes armadillos dig the holes close to houses. The digging can damage concrete and pipes.

Chipmunks dig holes.

Birds move sticks to make nests.

Animals drop seeds as they eat.

Animal Food Trails

Animals change the environment as they find food for themselves. Some animals eat the plants around them. They drop plant seeds as they move. Then, new plants grow in new places.

(t) Paul E Tessier/Getty Images; (b) altrendo nature/Getty Images

Big animals create paths as they move around to find food and water. Long ago, people followed these trails. Later, people built roads along these paths. Now, animals have to move away from these roads to find a better environment for themselves.

Animals make paths.

Discussion Questions

Fiction

- What was Megan's favorite part about visiting her Grandpa?
- What happened to the oak tree?
- What used to be in the empty lot?

Nonfiction

- How can plants help other plants, animals and people when it is too hot?
- What is one way an animal can change its environment?
- What is one way a plant can change its environment?